AMAZING METEORITES

Told and illustrated by Richard & Naira Matevosyan

From "Father Jaguar and the Cub Like to Travel" series

To incurable followers of the celestial cues.

Not so long ago, one starry night father jaguar and the cub took a tour of amazing meteor showers. According to the "Jaguar @ NASA" itinerary, the nearest

shower should be at its best the night of Saturday, October 20th, and Sunday, October 21st, 2012.

"The moon sets by around midnight," noticed the jaguar, "so it won't pollute the sky with light during the peak meteor-watching hours."

"Where is the spot?" asked the cub.

"The constellation Orion," the jaguar double-checked. "Let's see what our itinerary suggests:

Name	Date of Peak	Moon in 2012
Quadrantids	Night of January 3	Sets after midnight
Lyrids	Night of April 21	New
Eta Aquarids	Night of May 5	Full
Perseids	Night of August 11	Morning crescent
Orionids	Night of October 20	First quarter
Leonids	Night of November 17	Evening crescent
Geminids	Night of December 13	New

The *Orionid* is the most prolific meteor shower associated with Halley's Comet. It is an annual meteor shower which lasts approximately one week in late-October."

"What is the *Orionid*?" asked the cub.

"The *Orionids* are so-called because the point they appear to come from, called the radiant, lies in the constellation Orion. Look at the celestial map of the Orion the Hunter. *Torsten Bronger*, a talented physicist from Aachen (Germany), has created it on August 18, 2003, by using the program PP3.The map is calculated with the equidistant azimuthal projection (the zenith being in the center of the image). The north pole is to the top."

4

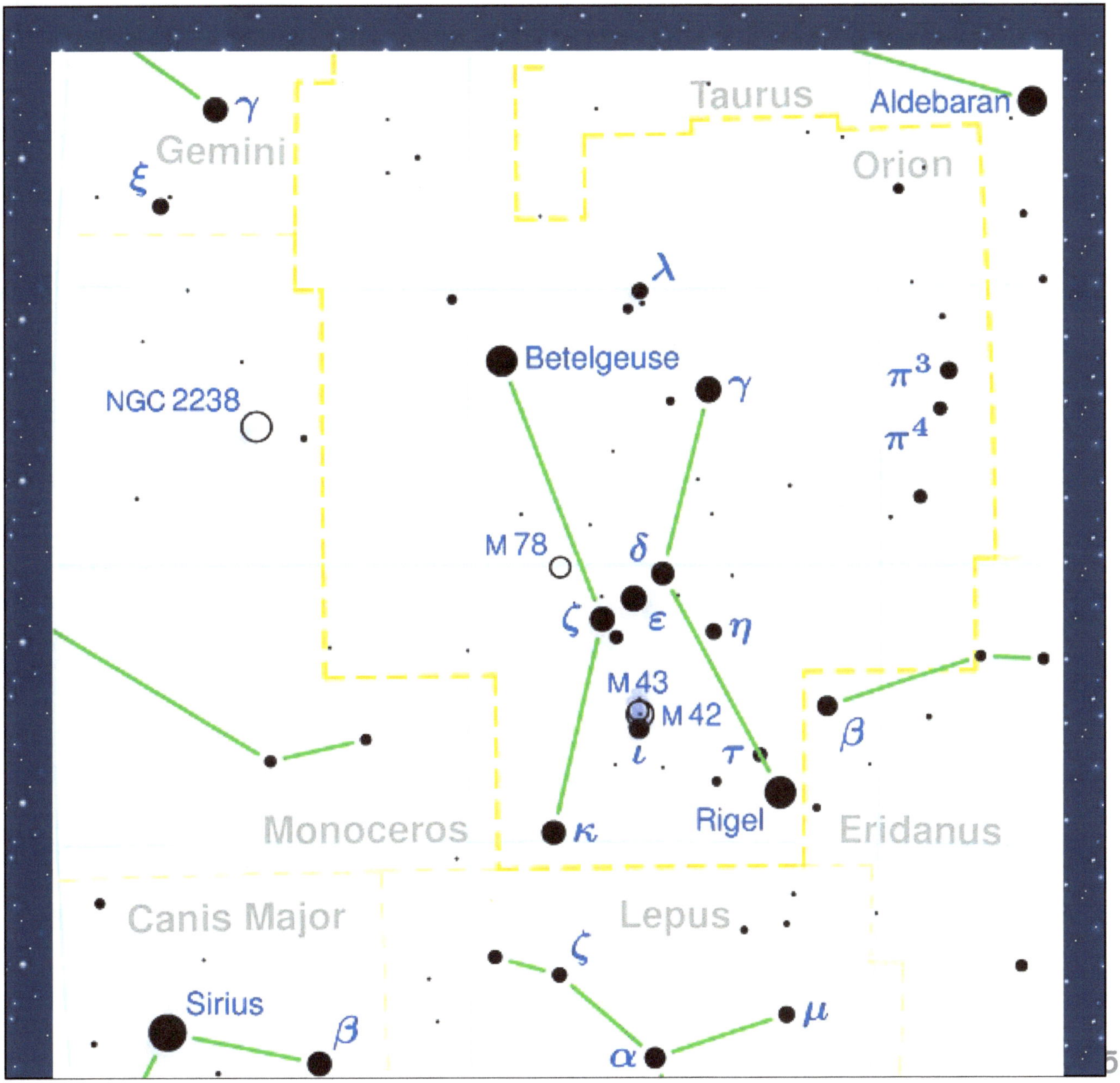

Taurus

Aldebaran

γ

Gemini

ξ

Orion

λ

Betelgeuse

γ

π³

π⁴

NGC 2238

M 78

δ

ζ **ε**

η

M 43

M 42

ι

τ

β

κ

Rigel

Eridanus

Monoceros

Canis Major

Lepus

ζ

Sirius

β

μ

α

"Dad, I still have no clue what are you talking about," meowed the cub.

"Just forgot! Your cubgarten does not run an astronomy class," sniffed the jaguar.

"That's true. We only study *Our land, Gravitational physics, Anatomy of battle, Intuition and timing, Healthy nutrition, Dentistry, Grassroots leadership,* and *Parasitology*," said the cub. "Now, tell me what a meteor is, ple-e-e-e-ase."

"A meteor is a bright streak of light (a shooting star) across the night sky produced by the entry of a small meteoroid into the Earth's atmosphere. The word "meteor" originates from the Greek *meteora* or *meteoros*, once used to describe any atmospheric occurrence, such as auroras, lightening, rainbows, and the like. Historically, the altitude at which meteors appeared was a subject of controversy. It was resolved by the fact that the location of a given meteor in the sky would appear to shift depending from where on the ground the meteor was seen.

This apparent positional shift is referred to as *parallax*. It wasn't until the 18th century that the heights of meteors were first calculated, using two observers at different locations and parallax. The approximate height of a meteor

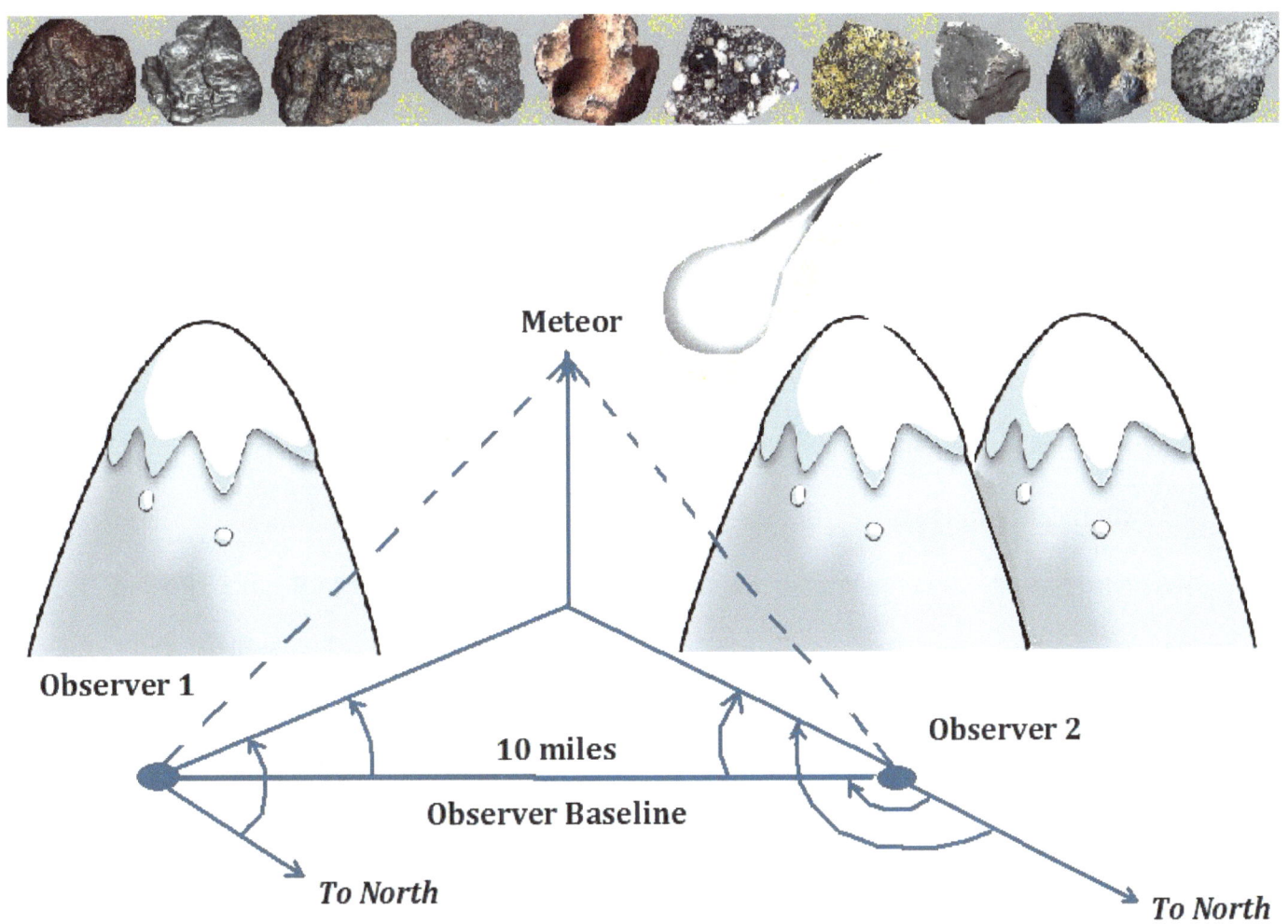

seen by two observers at two different locations were determined by using trigonometry. That information helped to predict the location of a possible meteorite fall. However, these initial calculations which showed that meteors were very low in the earth's atmosphere were incorrect, because the two points of

measurements were not far enough apart.

By the end of the 18th century, the accurate data collected by two German students, Henrich William Brendes and Johann Friedrich Benzenberg laid the foundations for subsequent studies concerning the altitude of most meteors in excess of 40 - 60 miles above the surface of the earth."

"What makes meteors?" asked the cub.

"Meteors are caused by small bits of interplanetary rock and debris called meteoroids vaporizing high in Earth's upper atmosphere. Traveling at tens of thousands of miles an hour, meteoroids quickly ignite from the searing friction with the atmosphere, 30 to 80 miles above the ground. Almost all are destroyed in this process; the rare few that survive and hit the ground are known as meteorites. If you have a dark clear sky you will probably see a few per hour on an average night; during one of the annual meteor showers you may see as many as 100/hour. Very bright meteors are known as fireballs or bolides.

Most meteor showers are spawned by comets. As a comet orbits the Sun it sheds an icy, dusty debris stream along its orbit. If Earth travels through this

stream, we will see a meteor shower. Although the meteors can appear anywhere in the sky, if you trace their paths, the meteors in each shower appear to "rain" into the sky from the same region.

When a meteor appears, it seems to "shoot" quickly across the sky, and its small size and intense brightness might make you think it is a star. If you're lucky enough to spot a meteorite and see where it hits, it's easy to think you just saw a star "fall." Most come from asteroids, including few are believed to have come specifically from 4 Vesta; a few probably come from comets. A small number of meteorites have been shown to be of Lunar or Martian origin. Many meteorites preserve chemical and physical properties that were established 4.5 billion years ago, during the earliest history of the solar system, and thus provide some of the best clues to the nature of the events that occurred in that remote time. One of the Martian meteorites, known as *ALH84001*, is believed to show evidence of early life on Mars!"

"Wow! Look at this! I am touching a piece of Mars!" exclaimed the cub.

"I know how it feels," roared the jaguar. "That's a momentum. Here is what

President Bill Clinton had to say about the discovery of life evidence in a meteorite

from Mars: *'It is well worth contemplating how we reached this moment of discovery. More than 4 billion years ago this piece of rock was formed as a part of the original crust of Mars. After billions of years it broke from the surface and began a 16 million year journey through space that would end here on Earth. It arrived in a meteor shower 13,000 years ago. And in 1984 an American scientist on an annual U.S. government mission to search for meteors on Antarctica picked it up and took it to be studied. Appropriately, it was the first rock to be picked up that*

year -rock number 84001. Today, rock 84001 speaks to us across all those billions of years and millions of miles. It speaks of the possibility of life. If this discovery is confirmed, it will surely be one of the most stunning insights into our universe that science has ever uncovered. Its implications are as far-reaching and awe-inspiring as can be imagined. Even as it promises answers to some of our oldest questions, it poses still others even more fundamental."

"That's thrilling! Something from the heaven just crushed your prey, and you're pretty sure it wasn't a campfire ball," the cub imagined.

"I know! A small meteorite may fall without any sound or light effects. The fall of a larger meteorite can be accompanied by startling light and sound effects. A fiery mass suddenly appears in the sky traveling swiftly in an arc and leaving a luminous trail behind. It may then disintegrate with a loud explosion and its fragments fall to the ground. Most meteorites break into pieces during the luminous flight and produce multiple individual fragments.

Meteorites are heated by friction with the air as they pass through Earth's upper atmosphere. Their outer surface is melted and continually removed by

airflow. Although the meteorite in flight appears incandescent, the heat generated in this process is mainly lost to the surrounding atmosphere, so little heat actually penetrates the cold meteorite. Sonic booms are frequently produced at this stage of flight. The energy resulting from the high velocity of entry into the atmosphere is dissipated within a few seconds while the meteorite is still at high altitude, and the body then falls freely and comparatively slowly to the ground. This long plunge through the cold atmosphere cools the meteorite considerably. Interestingly, meteorites do not ignite grass or fall in flames."

"What's the difference between the meteors and meteorites?"

"Say you're a bit of interplanetary dust or debris trucking through the vacuum of space, minding your own business. You're not very big. Certainly not big enough to be called an asteroid. In fact, you might just be a speck of dust or even smaller. Congrats! You're a *meteorite!* But say, for example, a bright blue planet suddenly gets in your way and sucks you in, and before you know it you're streaking through an atmosphere so fast that you ablate and let off a bright streak of light. You are now officially a *meteor.*

Now, on the other hand, if you started out big enough, then enough of you will emerge from this furnace o' friction to hit the ground in some farmer's field, making you a meteorite. Mazel tov! Of course, we should specify here: If you're made of rock, you're called a *chondrite*. Mostly metal? You're an *iron meteorite*. A little of both? Say, a rock wrapped in metal? You are hereby dubbed a *pallasite*. Below are the types of meteorites stratified for their chemical compounds:

Iron	*Stony Iron*	*Chondrite*	*Carbonaceous Chondrite*	*Achondrite*
Mixture of iron and nickel	Mixture of iron and stony material	Similar to the mantles and crusts of the terrestrial planets	Similar to the composition of the Sun less volatiles	Originated on the Moon and Mars, and similar to terrestrial basalts

Achondrites are 'high-energy' meteorites in the sense that they appear to have been chondritic before being altered by an immense heating or impact event at an asteroidal or planetary level. They are much rarer than chondrites and include

the *Eucrites* (asteroid Vesta), *SNC's* (abbreviations for the Mars originated Shergottite, Nahklite, and Chassiginite), *Howardites*, *Brachinites*, *Aubrites*, *Ureilites*, and *Diogenites*. Additionally, there are further groupings such as *Acapulcoites*, *Winonaites*, and *Lodranites*.

"And what is an asteroid?" the cub asked.

"Just as meteoroids are too small to be called asteroids, asteroids are too small to be called planets. Most asteroids that have been discovered (now well over 100,000) range from about 10 kilometers to 100 kilometers in diameter and are found in a belt between Mars and Jupiter. Scientists believe that this asteroid belt

was a nascent planet that bullied by Jupiter's immense gravity, never quite got it together. Asteroids are grouped into a number of families based on their orbits. Those whose orbits cross the Earth's orbits are called Apollo asteroids (see above).

Earth crossing is not a good thing, however. If an asteroid just 1 kilometer across were to hit the Earth, we wouldn't get a pretty light show. We'd get a jolt equivalent to a 20-megaton nuclear blast, leaving a crater with a diameter equal to the length of Manhattan. That kind of jolt is survivable—if you're a character in the movie *Deep Impact*. If you're a jaguar on Earth, it stirs up a bit more

trouble no matter how you can jump."

"Are there any traces we know?" asked the horrified cub.

"You bet! A good example of what happens when a small asteroid hits the Earth is *Barringer Crater* near Winslow, Arizona (see above). Formed about 50,000 years ago by an iron meteor about 30-50 meters in diameter, 1200 meters in diameter and 200 meters deep.

About 120 impact craters have been identified on the Earth. The impact of a comet or asteroid about the size of *Hephaistos* or *SL9* hitting the Earth was

probably responsible for the extinction of the dinosaurs 65 million years ago. It left a 180 km crater now buried below the jungle near *Chicxulub* in the Yucatan Peninsula.

Another impact occurred in 1908 in a remote uninhabited region of western Siberia known

as *Tunguska*. The impactor was about 60 meters in diameter, consisted of many loosely bound pieces. In contrast to the *Barringer Crater* event, *Tunguska* completely disintegrated before hitting the ground and so no crater was formed. Nevertheless, all the trees were flattened in an area 50 kilometers across. The

sound of the explosion was heard half-way around the world in London."

"How *fall* and *find* are related to these shows?" the petrified cub opened the claws.

"A "fall" means the meteorite was witnessed to fall from the sky. A "find" means the meteorite was found at a later time," explained father jaguar.

"Who cares of the difference?" the cub tried to see the point.

"Meteorite fall statistics are frequently used by planetary scientists to approximate the true flux of meteorites on the Earth. Although there are 30 times more *finds* than *falls*, their raw distribution of types does not accurately reflect what falls to Earth.

The following table is from a book by *Vagn F. Buchwald*. Included are all known meteorites (4,660 in all, weighing a total of 494,625 kg) in the period 1740-1990 (excluding meteorites found in Antarctica).

Type	Fall	Find	Fall weight (kg)	Find weight (kg)
Stony	95.0 %	79.8 %	15, 200	8, 300
Stony-Iron	1.0 %	1.6 %	525	8, 600
Iron	4.0 %	18.6 %	27, 000	435, 000

"What about the chondrite/achondrite statistics ?" meowed the cub.

"For some of the less-studied stony meteorite falls, it is not known whether the object is chondritic or achondritic. Overall, the annual chondrite/achondrite fall ratio equals 10.6 (for 915 and 86, correspondingly)," concluded the jaguar.

"Poor planet Earth!" deplored the cub. "How long it suppose to take the celestial slaps?"

"Tell me about that! Believe it or not, about 25 million meteoroids hit Earth's atmosphere every day. And while most of them burn away to nothing, sometimes the Earth's orbit will take them through a messy patch of interplanetary waste, like the orbit of a dead comet that's broken into millions of meteoroids. In such a case, the Earth's gravity can hoover up these particles by the millions—creating meteor showers. A huge shower *Perseid* emanates every August for instance, creating an event that's widely publicized and not to be missed."

"*Perseid?*" asked the cub.

"Meteor showers are named for the constellation that coincides with this region in the sky, a spot known as the radiant. The *Perseid* meteor shower is so

named because meteors appear to fall from a point in the constellation *Perseus*,"

yawned the jaguar.

"How can I best view a meteor shower?" the cub licked his dad, giving him a hard time to sleep.

"Well, buddy! Let's look again at our itinerary. We suppose to get away from the glow of city lights and toward the constellation from which the meteors will appear to radiate. For example, we have to head the north to view the *Leonids*, our favorite. Jumping south may lead us to darker skies, but the glow will dominate the northern horizon, where *Leo* rises. *Perseid* meteors will appear to "rain" into the atmosphere from the constellation *Perseus*, which rises in the northeast around 11 pm. in mid-August. After we've escaped the city glow, we have to find a dark, secluded spot where oncoming car headlights will not periodically ruin your sensitive night vision.

In 2012, the full moon gets in the way of the May *Eta Aquarids*. Moon-free nights greet the April *Lyrids*, the November *North Taurids* and the December *Geminids*. Moonlight should not pose much of a problem for the October *Draconids*, October *Orionids*, November *South Taurids* and November *Leonids*. Some moon-

free viewing time is in store for the January *Quadrantids* and July *Delta Aquarids*."

"Sounds great! Once we settle at our observing spot - far from people, we shall lie back to position, so the horizon appears at the edge of our peripheral vision with the stars and sky filling our field of view. Meteors will instantly grab my attention as they streak by," suggested the cub. "What shall we pack for meteor watching?"

"Treat meteor watching like you would with the fireworks on 4th of July. Seat comfortable, pack food, water, plus a red-filtered flashlight for reading maps and charts without ruining your night vision. Binoculars are not necessary. Your cat-eyes will do just fine!"

"Like I said, October 2012 is rich in 'shows.' In fact, there are two annual meteor showers in October that give us a chance to see *shooting stars*. The *Draconids* around nightfall and early evening on October 7 and 8, can be watched from any part of the globe. The radiant point for the *Draconid* shower almost coincides with the head of the constellation *Draco* the Dragon in the northern sky. That's why the *Draconids* are best viewed from the Northern Hemisphere. The

Draconid shower is a real oddity, in that the radiant point stands highest in the sky as darkness falls. Unlike many meteor showers, the Draconids are more likely to fly in the evening hours than in the morning hours after midnight. This shower is

Deneb

Head of Draco

Summer Triangle

Vega

Altair

Southeast

Southwest

West

usually a sleeper, producing only a handful of languid meteors per hour in most years. But watch out if the Dragon awakes! In rare instances, fiery *Draco* has been known to spew forth many hundreds of meteors in a single hour. With no moon to interfere during the evening hours, try watching at nightfall and early evening on October 7 and 8."

"Then comes the *Orionid*," noticed the cub, "before dawn on October 21-22."

"That's correct! The discovery of the *Orionid* shower should be credited to E. C. Herrick (Connecticut, U.S). The waxing crescent moon setting before midnight (on October 20th) secures the best viewing hours for the *Orionid* showers. On a dark, moonless night the *Orionids* exhibit a maximum of about 15 meteors per hour. These fast-moving meteors occasionally leave persistent trains and bright fireballs. You might know the constellation Orion's bright, ruddy star *Betelgeuse*. The radiant is north of *Betelgeuse*. The *Orionids* have a broad and irregular peak that isn't easy to predict. More meteors tend to fly after midnight, and the *Orionids* are typically at their best in the wee hours before dawn."

Orion

radiant

Sirius
brightest star in the sky

East

Southeast

Orionid Radiant

Late night of November 4th until dawn November 5th is the big time of
South Taurids. The South and North *Taurids* are perhaps best suited to die-hard

meteor aficionados. The meteoroid stream that feeds the *Taurids* is very spread out and dissipated. That means the *Taurids* are extremely long lasting (September 25 - November 25) but usually don't offer more than about 7 meteors per hour. The waxing crescent moon sets at early evening, leaving a dark sky for the *South Taurid* meteors, which are expected to produce the most meteors in the wee hours after

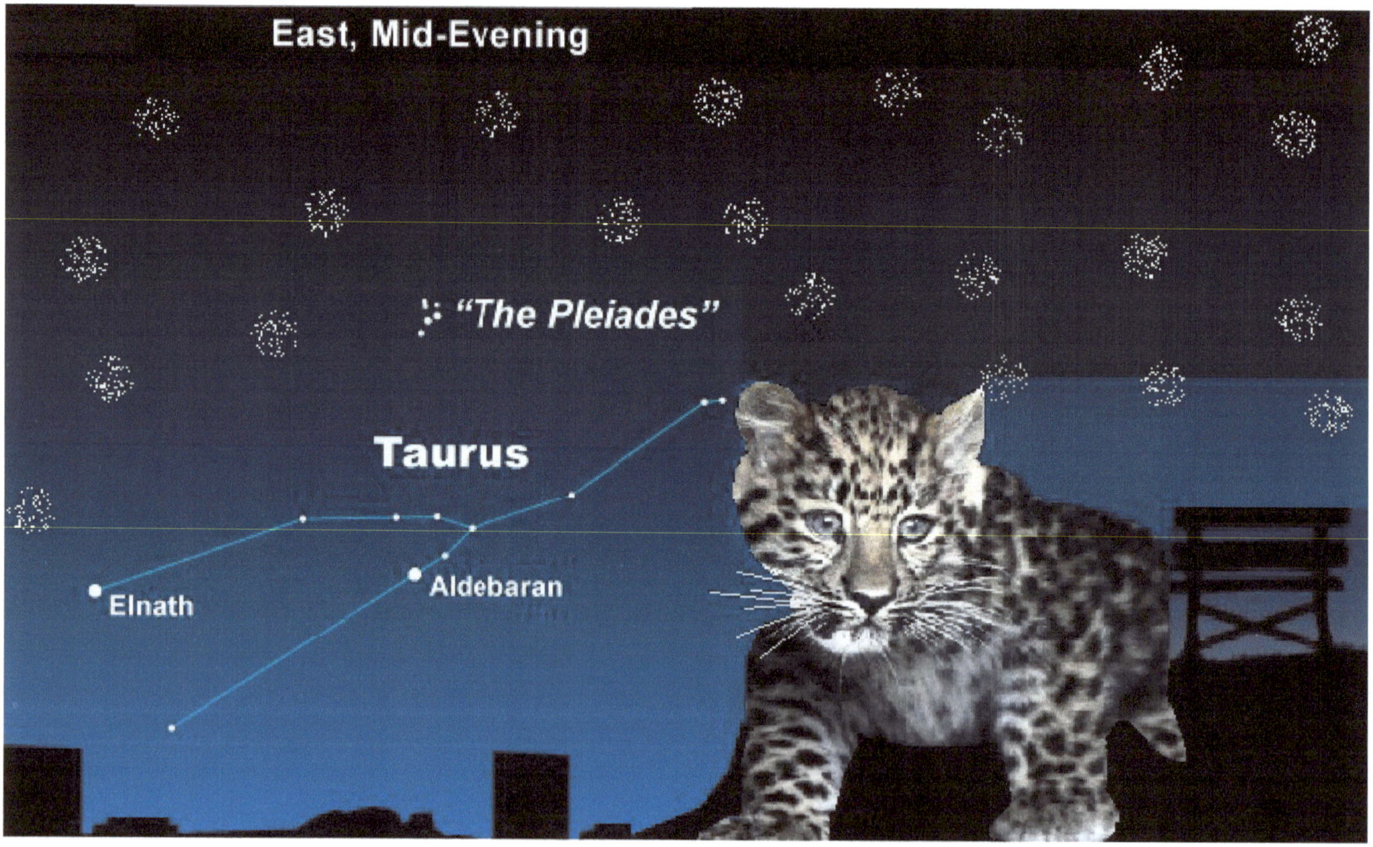

East, Mid-Evening

☆ *"The Pleiades"*

Taurus

Aldebaran

Elnath

midnight on November 5th."

"What comes next?" the cub looked at the itinerary. "*North Taurids!* We must watch it in the late night November 11th until dawn November 12th, 2012. This shower is long-lasting (October 12 – December 2) but modest, and the peak number is forecast at about 7 meteors per hour. Typically, we will see the maximum numbers at around midnight to 1 am, when *Taurus* the Bull moves nearly overhead. This year, the thin waning crescent moon won't rise till close to dawn, leaving a long dark night for these rather slow-moving but sometimes bright *North Taurid* meteors. We will see some *Taurid* fireballs. I can't wait to savor the dazzling moment!"

"Bear in mind, that the bright moonlight of November can steal the show," roared father jaguar. "The bright waxing gibbous moon usually outshine the *South Taurids* shower on November 5th. The *North Taurids* might be tricky to spot, for these slow-moving meteors amid the bright moonlight. North Taurids are weaker. Taking into account the increasingly bright moonlight, the *Taurids* should be visible to skywatchers in North America with clear skies."

"The next is the *Leonids*, our family!" the proud cub raised the tale. "The annual Leonid meteor shower is expected to peak the night of November 17th."

"The waxing crescent moon setting at early evening may be a major interference, and we could see a rate of about 20 per hour," the jaguar suggested. "While the *Leonids* have typically been one of the more stunning light shows of the year, both showers could be more subdued this year because of a bright moon. Particles may encounter Earth on November 16th at around 5:30 pm EST, where we could see anywhere from 100 to 200 meteors per hour. We could get a *Leonid* outburst, but unfortunately it is not favorably placed for viewing from the U.S."

"So, then November 17th!" confirmed the cub.

"Precisely so. Radiating from the constellation *Leo* the Lion, the *Leonid* meteor shower is famous. Historically, this shower has produced some of the greatest meteor storms in history – at least one in living memory, 1966 – with rates as high as many thousands of meteors per hour. Indeed, on that beautiful night in 1966, the meteors did fall like rain. Some who watched the shower said they felt as if they needed to grip the ground, so strong was the impression of Earth plowing

along through space, fording the meteoroid stream. The meteors after all, were all streaming from a single point in the sky – the radiant point – in this case in the constellation *Leo*. *Leonid* meteor storms sometimes recur in cycles of 33 to 34 years, but the *Leonids* around the turn of the century – while wonderful for many observers – did not match the shower of 1966."

"The guide suggests that the *Leonids* in particular are well known for having bright meteors which may be 9 mm across and have 85 g of mass and punch into the atmosphere with the kinetic energy of a car hitting at 60 mph. An annual Leonid shower may deposit 12 or 13 tons of particles across the entire planet. The meteoroids left by the comet are differentially disturbed by the planets, in particular Jupiter and to a lesser extent by radiation pressure from the Sun, the Poynting–Robertson effect, and the Yarkovsky effect. Please explain!"

"The *Poynting–Robertson effect* is a process by which solar radiation causes a dust grain in the Solar System to slowly spiral into the Sun. The *Yarkovsky effect* is a force acting on a rotating body in space caused by the anisotropic emission of thermal photons, which carry momentum. In most years, the *Lion* whimpers rather than roars, producing a maximum of perhaps 10-15 meteors per hour. Like most meteor showers, the *Leonids* ordinarily pick up steam after midnight and display the greatest meteor numbers just before dawn."

"The December's shower is 'sponsored' by the *Geminid* , often producing 50 or more meteors per hour," meowed the cub.

"It is a beloved show, because as a general rule, it's either the August *Perseids* or the December *Geminids* that give us the most prolific display of the year. Best of all, the new moon guarantees a dark sky on the peak night of the *Geminid* shower (mid-evening December 13th until dawn December 14th). But the nights on either side of the peak date should be good as well," explained the jaguar.

"Unlike many meteor showers, you can start watching the *Geminids* by 9pm or 10pm local time. The peak might be around 2 am, because that's when the shower's radiant point is highest in the sky as seen around the world. With no moon to ruin the show, 2012 presents a most favorable year for watching the grand finale of the annual meteor showers!"

"Caused by the object 3200 Phaethon (a Palladian asteroid), the *Geminids* are

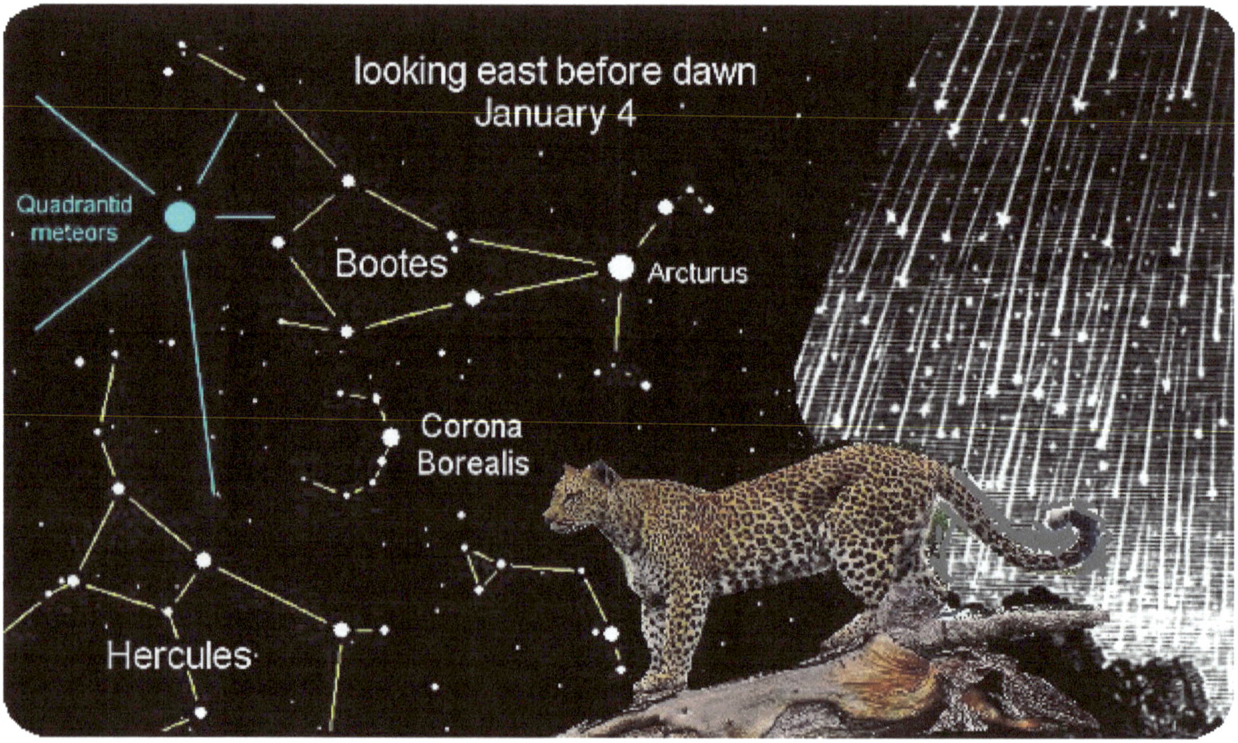

looking east before dawn
January 4

Quadrantid meteors

Bootes

Arcturus

Corona Borealis

Hercules

the stunning comet-originated pre-Christmas gloomy showers," excelled the cub.

"Tell me about it!" roared the jaguar. "With the arrival of the next year, we admire another miracle, the *Quadrantids*. Although the *Quadrantids* can produce over 100 meteors per hour, the sharp peak only lasts for a few hours, and doesn't always come at an opportune time. In other words, you have to be in the right spot on Earth to view this shower in all its splendor. Speaking of January 4th, it means to be the wee hours before dawn, not the night. Although the waxing gibbous moon lights up most of the night and doesn't set until roughly 3 am, this is about the best time to watch for this shower. Usually, eastern North America, the North Atlantic Ocean and possibly western Europe are in a fine position to watch this show. The *Quadrantid* is worth a try at northerly latitudes all around the globe."

"April 22nd, a big time for the *Lyrids*!" noticed the cub.

"The *Lyrid* meteors, named April's "shooting stars" tend to be bright and often leave trails. About 10-20 meteors per hour at peak can be expected. Plus, the *Lyrids* are known for uncommon surges that can sometimes bring the rate up to 100 per hour. Those rare outbursts are not easy to predict, but they're one of the

reasons the tantalizing *Lyrids* are worth checking out. The radiant for this shower is in the constellation *Lyra*, which rises in the northeast at about 10 pm. In April 2012, the new moon guaranteed a dark sky in the late night and morning hours, the

Radiant

Vega

LYRA

Deneb

CYGNUS

AQUILA

Altair

10°

April 2 A.M.
Looking east

best time to watch the *Lyrid* shower. As a general rule, the greatest number of *Lyrid* meteors fall in the dark hours before dawn," explained father jaguar.

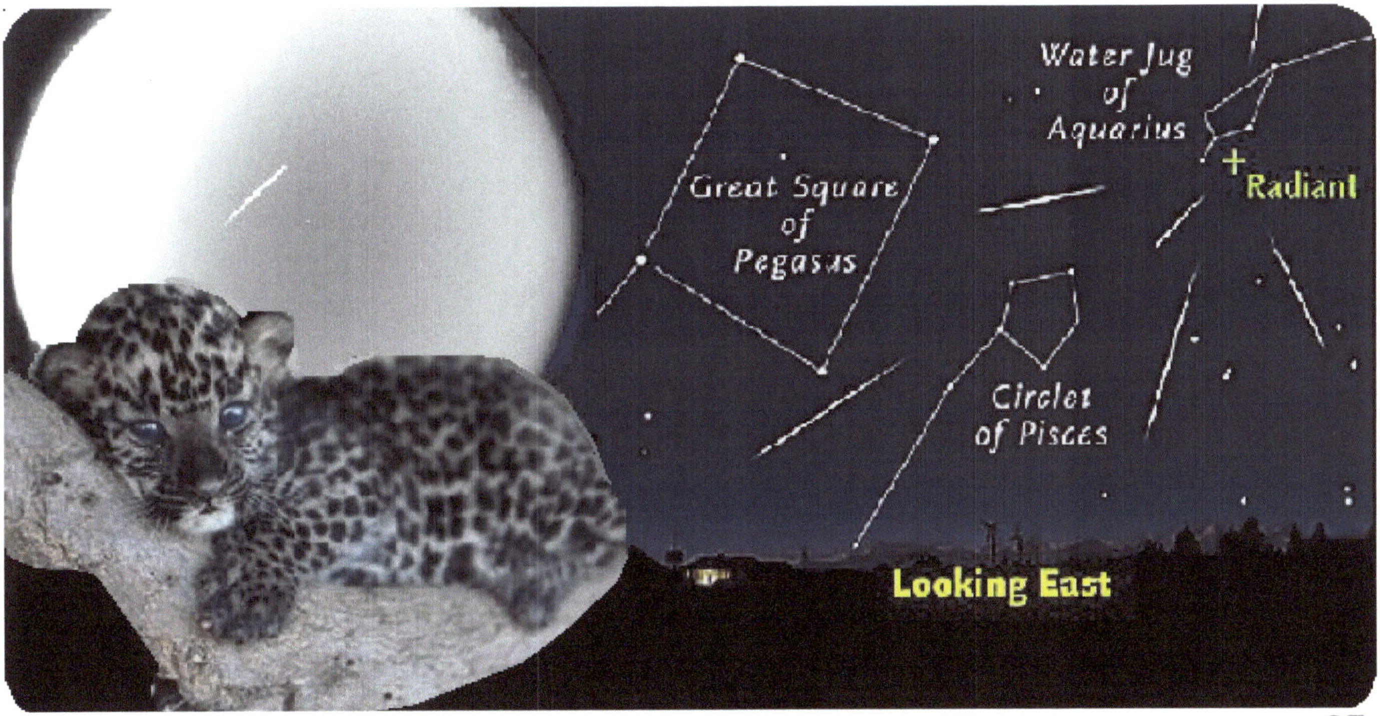

"*Eta Aquarids* shower has a relatively broad maximum but is expected to show the greatest number of meteors before dawn on May 5th or 6th. This year (2012), the closest and largest full moon was out all night long, leaving no dark sky for the grand show. But if were a die-hard meteor enthusiasts you would see the shower in a moonlit sky. In the northern United States, Canada or Europe – the meteor

numbers are few and far between. In the southern half of the United States 10 to 20 meteors per hour might be visible in a dark sky. In the Southern Hemisphere the meteor numbers increase dramatically. Perhaps two to three times more *Eta Aquarid* meteors streaking the southern skies. For the most part, this is a predawn shower. The radiant for this shower appears in the east-southeast at about 4 am and the hour or two before dawn offers the most meteors," suggested the jaguar.

Aquarius

S. Delta Aquarids
2-4 a.m. July 28-29

Delta

Fomalhaut

South

"I see. So, the full moon is washing away all but the brightest *Eta Aquarid* meteors," clarified the cub.

"It sure does," yawned father jaguar.

"Like the Eta Aquarids, *Delta Aquarids* shower favors the Southern Hemisphere, and the tropical latitudes in the Northern. Although the waxing gibbous moon won't set till after midnight, the hours between moonset and dawn offer the most *Delta Aquarid* meteors radiating from the southern part of the sky. From northern temperate latitudes, the maximum hourly rate may reach 15-20 meteors in the dark. Unlike many showers, this one doesn't have a very definite peak. Instead, these medium-speed meteors ramble along fairly steadily throughout late July and early August, between moonset and dawn," noticed the jaguar.

"The *Perseids* are typically fast and bright meteors. They radiate from a point in the constellation *Perseus* the Hero. Each year in August 10-12, with the meteors in late evening as well. On any of those mornings, moonlight will be shouldn't be so overwhelming as to ruin the show.

Plus the moon on those mornings near the bright planets Venus and Jupiter in the eastern predawn sky. It'll be a moon rising into the predawn sky, we watch for Perseid beautiful early morning scene. You don't need to know Perseus to watch the shower because the meteors appear in all parts of the sky," roared the jaguar. "The *Perseids* are

considered by many to be the year's best shower, and often peak at 50 or more meteors per hour in a dark sky. They streak across these short summer nights from late night until dawn, with only a little interference from the waning crescent moon."

"The tour of the amazing showers is over and I have so many questions: why the meteor showers happen? What cause them, and why they occur in certain patterns? Why they are predictable?" the cub kept asking.

"There are two kinds of meteors – *sporadic* and *shower* meteors. Sporadics originate from random bits of solar system dust that orbit the Sun. Their chance encounters with Earth are unpredictable. While they do slightly cluster in various parts of the sky, their occurrence is sporadic – hence the name. Sporadics are the ones most people see while gazing into the night sky. Naked-eye rates for sporadic meteors seldom exceed five per hour. As far as we know, all meteors that reach the ground - meteorites - come from sporadics.

Shower meteors come from the dust released by comets as they travel through our solar system. The dust spreads out along the comet's orbit and forms

an elliptical trail of debris that passes around the Sun and crosses the orbits of the planets. Meteor showers occur when the Earth passes through this trail of debris during its yearly orbit around the Sun. The following year, Earth passes through that same debris trail again on about the same date. This is why meteor showers are predictable annual events. Now, look at the picture," the jaguar licked the cub's face. "In this diagram you are looking down onto the Earth's North Pole. Note how the morning side of Earth will plow into the dust but the evening side will be somewhat shielded. This is

why there are often more visible meteors after midnight - you are then on the side of the Earth that is plowing into the dust."

"How do comets produce meteor showers?" asked the cub.

"Comets are small bodies composed primarily of ice with a little bit of sand or

gravel. A typical comet's nucleus is a few miles across. It spends most of its time in a lazy, elliptical orbit in the outer solar system where its nucleus is cold and largely inactive. For example, Halley's comet has a period of 76 years and at its furthest point from the sun is beyond the orbit of Neptune. Here the surface temperature of the comet is about 47 degrees above absolute zero (-375 F). But during the comet's passage near the Sun, its surface heats up, some of the ice evaporates and dust is released. Each comet has two tails, one composed of dust, the other of gas. Both tails stretch away from the nucleus and point more or less away from the Sun. This is because very hot particles coming from the Sun (solar wind) push the tails outward, regardless of the direction the nucleus is moving.

The dust streams may look uniform but they usually consist of several individual streams, like strands of a rope. Each strand was produced by a different passage of the comet through the inner solar system. The elliptical stream of particles also shifts very slightly from year to year owing to Jupiter's gravitational field. As a result, the number of meteors can vary from one annual shower to the next as the Earth passes through different parts of the dust stream. By about

2099 the orbit of comet *Temple-Tuttle* (source of the *Leonid* meteors) will no longer intersect the Earth's orbit. The result? No more *Leonid* meteor shower."

"So, comets are the origin of most meteor showers but a few come from asteroids," concluded the cub.

"These may well be very old comets. After enough passages through the warm, inner solar system, the ice has been completely evaporated, leaving a loose assemblage of dust particles held together by their own feeble gravity. These "rubble pile" asteroids may be the remains of former comets," said the jaguar.

"What is the physical-chemical composition of the meteors?" the cub was digging the sand.

"Meteors produce hot trails of ionized gas behind them. Some of these trails may be visible in the night sky for several minutes after the meteor passes. This gas reflects radar waves. As a result the meteors are also detected in daytime."

"Are there any man-made meteors we should know?" the cub stretched the claws.

"Nowadays, some meteors are actually bits of man-made space debris. These

tend to be things like paint chips and spent rocket hardware. The meteors they produce can sometimes be identified as man-made because they travel much more slowly across the sky than natural meteors," explained the jaguar.

"Dad. The showers are over. It's time to find meteorites!" the cub was thrilled. "How we suppose to detect them?"

"Humans often think of meteorites as heavy and magnetic rocks, with a shiny, black fusion crust and flow marks from melting. They often confuse them with the terrestrial rocks covered in desert varnish, slag or iron ores such as magnetite or hematite. Probably 99% of the items guessed as meteorites turn out to not be meteorites," the jaguar noticed.

"Isn't it true that all meteorites are highly magnetic and won't mark a streak plate?" asked the cub.

"No. While it's true that freshly-fallen meteorites are usually magnetic and

won't mark a streak plate, the most common cold finds are weathered ordinary chondrites, like this one, which of are often only weakly magnetic because of the oxidation (rusting) the iron they contain. Because

of the oxidation, they'll leave a brownish- orange streak on an unfinished piece of ceramic," the jaguar jumped and caught a piece of chondrite. "The most common magnetic "meteorwrong" that won't mark a streak plate is slag. Magnetite, though highly magnetic, will leave a gray streak, and hematite will streak reddish-brown.

"But don't all meteorites have a shiny, black fusion crust?" asked the cub.

"The longer a meteorite has been on Earth, the more the fusion crust wears away, leaving the meteorite a rusty brown color. This process is accelerated in humid areas and regions that receive large amounts of precipitation," explained father jaguar." As for the fusion crust, it is a layer that forms on the outer

Magnetite

Hematite

surface of a meteorite as the result of frictional heating and abrasion as the meteorite enters the Earth's atmosphere. Freshly-fallen meteorites usually have a shiny, black fusion crust.

Keep in mind that none of the descriptive features is 100% conclusive. Terrestrial (earth) features that indicate a rock is not a meteorite. If there is quartz (a clear or milky white crystal) or a transparent crystal, it is not a meteorite. If the rock has many small holes in its interior it is probably not a meteorite. Small holes are almost never found in the interior of meteorites. If the rock feels light it is not a meteorite. Stony meteorites generally have a density around 3.5 grams per cm^3 and normal earth crustal rock is around 2.7 grams per cm^3. In other words stony meteorites are about 30% heavier for the same size rock. A rock that is denser than normal does not have to be a meteorite. There are many examples of dense terrestrial rocks. If the exterior has sharp pointed features (except on obviously broken surfaces) it is not a meteorite. As meteorites come through the atmosphere any sharp points are melted away. As a result meteorites have smooth surfaces, although they may have depressions, called

regmaglypts that look like thumbprints pushed into clay.

The clue to meteorites is foremost in the fusion crust. But it is one of the more difficult things for a novice to get a good feeling for. Different types of meteorites can have different crusts. The *Aubrite* called Cumberland Falls, has a very thin almost yellowish crust; whereas, the *Howardite*, Great Sand Seas 010, has a shiny black, almost plastic looking crust. It also has flow lines where melted crust flowed in streams pushed by the atmosphere during the entry:

Aubrite

Howardite

"Let's do a little homework, dad. We have to uncover meteorites among the rocks I collected, "the cub moved with the tale.

"We gonna run a *streak test,* " roared the jaguar." It involves scratching a sample across an unglazed ceramic tile to see if it leaves a mark. While freshly-

fallen meteorites won't mark a streak plate, the overwhelming majority of meteorite finds are weathered ordinary chondrites, which may streak brownish-orange. Hematite leaves a red-brown streak and magnetite leaves a gray-black streak. Other minerals may leave brown, black, green-black, gray, or even yellow streaks. If your specimen does not leave a streak, you may have a piece of slag: a man-made industrial byproduct of the mining and metallurgy industries.

Magnetite (Fe_3O_4) and hematite (Fe_2O_3) are two common meteorwrongs iron oxide minerals that attract magnets. Slag, a man-made byproduct of mining and metallurgy, is often made up of metal, sometimes combined with metal oxides or sulfides, and many additional components (silica, calcium, etc). Because slag is formed by the cooling of melted industrial byproducts, it often displays melt texture, such as flow marks and vesicles (holes or "bubbles" in the surface where trapped gas has escaped during cooling) and can be heavy and magnetic. It may even appear similar to some meteorites, so be wary of this meteorite impostor!"

"I guess slag can be found almost anywhere," said the cub, "even in what might be considered *'the middle of nowhere,'* because it is commonly used for fill in

roads or train tracks. Now tell me please: do meteorites attract a magnet?"

"Not necessarily. About 90% of iron and iron-stone meteorites will attract a magnet. If your rock does attract a magnet, its probably not a meteorite because magnetite-rich Earth rocks are much more common than meteorites. Cut or break it open. If it has lots of metal flecks or veins like these ordinary chondrites, then it might be a meteorite (but industrial slags sometimes contain metal)," explained the jaguar. "Some meteorites even contain gemstones. The beautiful *Brenham pallasite*, found in Kiowa County, Kansas is packed with sea-green olive crystals, which is also known as the semi-precious gemstone *peridot*. Both the *Allende* meteorite which fell in Chihuahua, Mexico, and the *Canyon Diablo* iron which formed Arizona's immense and erroneously named Meteor Crater contain micro diamonds. A few meteorites (*Murchison*) have been found to contain water and amino acids, compounds of carbon, hydrogen, and oxygen. These may be samples of the material that started our oceans and atmosphere, and perhaps provided the material from which life evolved.

The rarity of meteorites, along with the fact that they are the only way in

which most of us will ever have the chance to touch a piece of an alien world, make them of great interest to an ever-expanding network of private meteorite collectors."

"I agree. Meteorite collecting is an exciting and growing hobby and there are perhaps a thousand active enthusiasts in the world today. The international space rock market is something else we will explore in the months ahead," offered the cute and smart cub. "Now let is wrap up with the lesson. For testing the suspected meteorites I need to answer the following six questions:

(1) Does my rock have a dark-colored (typically black) thin exterior coating that shows evidence of melting and is clearly different from the light colored interior?

(2) Is my sample round?

(3) Is it very spongy (containing numerous holes)?

(4) Is the sample unusually heavy?

(5) Does it differ from the rocks typically found in the target area (strewn fields)?

(6) Does the sample attract a magnet?"

"Indeed. But after all, it is still hard to declare that the piece you hold in your paws is truly a celestial 'tourist.' There is one more important factor to consider in detecting meteorites," commented the jaguar. "It is the size of a territory you are able to cover. One can find a hundred coins on an acre of the park. In case of meteorite hunting, the number of meteorite finds is much lower - less than one per acre, excluding the areas where the meteorite showers are frequent. In order to cover a vast territory, you must have a search coil of a size that is bigger than a size of the conventional search coils. But the search coil's size should not be too small or too big. The search coil of 10.5"-15" in diameter is optimal."

"Got it!" jumped the cub. "A contemporary question for you: how much a meteorite costs?"

"The cost generally depends upon rarity and weight. In many instances, only a small amount of a particular meteorite is recovered. If your meteorite is of a rare classification, its cost will be greater than many pieces of a more common type. More often however, many common or rare meteorites are simply not available to

collectors and educators at any price!" sniffed the jaguar.

"Then, what shall we do with our hunted meteorites? This one is cool! It's a big stony-iron Imilac (Pallasite):"

"Let's report them to the Smithsonian Institution. As a memorabilia, we shall ask there to have our pictures taken together with our gifts," father jaguar and the cub ended the tour for the meteorite showers.

A 'FALLING MESSAGE' FOR THE JUNIOR READER

Many stories of bravery, romance, and tragedy are sedimented in us through our trends to comprehend the logic behind the celestial patterns. The intriguing phenomena of the night sky are tempered by intimidation but meantime in the held of reverence of the apparitions observed.

Current classification of meteorites suggests that there is a gap between and within different meteorite families. Taking into account meteoroid measurements by *in situ* experiments, zodiacal light observations, and oblique angle hypervelocity

impact studies, it is found that the observed size distributions of lunar microcraters usually do not represent the interplanetary meteoroid flux for particles with masses of 10^{-10}g. With many more meteorites waiting to be found in deserts and strewn fields, it is likely that we shall uncover distinct new types, to the extent of tiny monocomponent interstellar grains composed of a few tens of atoms of carbon, nitrogen, iron, manganese, silicon, or radium.

With predictive knowledge in meteor showers, the 'senders' and patterns, the azimuth and timing, the 'ambassadors' bringing 'coded messages' to our planet, we would better understand the stability of our solar system, refine the chronology of its tricks and miracles, and develop more robust timescales to date processes such as accretion, or parent-body processes, like aqueous alteration, metamorphism, and differentiation.

Of the books about the meteorites are apparently no end, but this particular effort is an oddity with its design to entertain and persuade the junior reader. Through a tour with avid travelers - father jaguar and the cub, it lands us on colorful scenes to specter the annual meteor showers, accommodating to the right

lunar phase (so the moonlight would not steal the show), geographical location, and the month. Witnessing the fall of the 'shooting stars,' we then find odd and regular meteorites with the help of various quests and tests.

Many postulate sources have suggested that the world ends in December 2012. This opinion claimed that Nibiru, a supposed planet discovered by the Sumerians, was headed toward Earth. This catastrophe was initially predicted for May 2003, but nothing happened. Then the fable was linked to the end of one of the cycles in the ancient Mayan calendar at the winter solstice in 2012 - hence the predicted doomsday date of December 21, 2012.

Nibiru and other stories about wayward planets have no factual basis. If Nibiru was real, astronomers would have been tracking it for at least the past decade, and it would be visible by now to the naked eye. As for the impacts by comets and asteroids, although big hits are very rare. The last big impact was 65 million years ago, and that led to the extinction of the dinosaurs. NASA's Spaceguard Survey is to find any large near-Earth asteroids long before they hit. Yet, no threatening asteroids as large 'to kill dinosaurs' are determined. The spatial

density of the detected meteoroids is presently increasing with the time. With the wealth of knowledge in these 'ambassadors' we would define the reason our lifesome planet was created, the path it goes through, and the fate it complies.

REFERENCES:

- **Allison R.J, McDonnell J.A.M (1981).** *Secondary cratering-effects on lunar microterrain: Implications for the micrometeoroid flux.* Proceedings of Lunar and Planetary Science (A conference presentation): 12B: 1703–1716

- **Bjorkman J.K (2012).** *Meteors and meteorites in the ancient near east.* Meteoritics and Planetary Science; 8 (2): 91-130

- **Bobrovnikoff N.T (1928).** *Meteors and meteorites.* Astronomical Society of the Pacific Leaflets; 1 (16):63

- **Grün E, Zook H.A, Fechtig H, Giese R.H (1985).** *Collisional balance of the meteoritic complex.* Icarus; 62 (2): 244-272

- **Reynolds M.D (2010).** *Falling Stars: A Guide to Meteors and Meteorites.* Stackpole Books, PA, 2nd edition (Paperback)

- **Norton O.R, Chitwood L (2008)**. *Field guide to meteors and meteorites.* Springer-Verlag, London Limited (Paperback)

- **Smith C, Russell S, Benedix G (2010).** *Meteorites.* A Firefly Book, Inc (Paperback)

- **Vogt G (2002)**. *Meteors and meteorites.* Capstone Press (Hardcover).